SCIENTIFIC AMERICAN EDUCATIONAL PUBLISHING

SCIENTIFIC AMERICAN INVESTIGATES THE HUMAN BODY

THE MUSCULAR SYSTEM

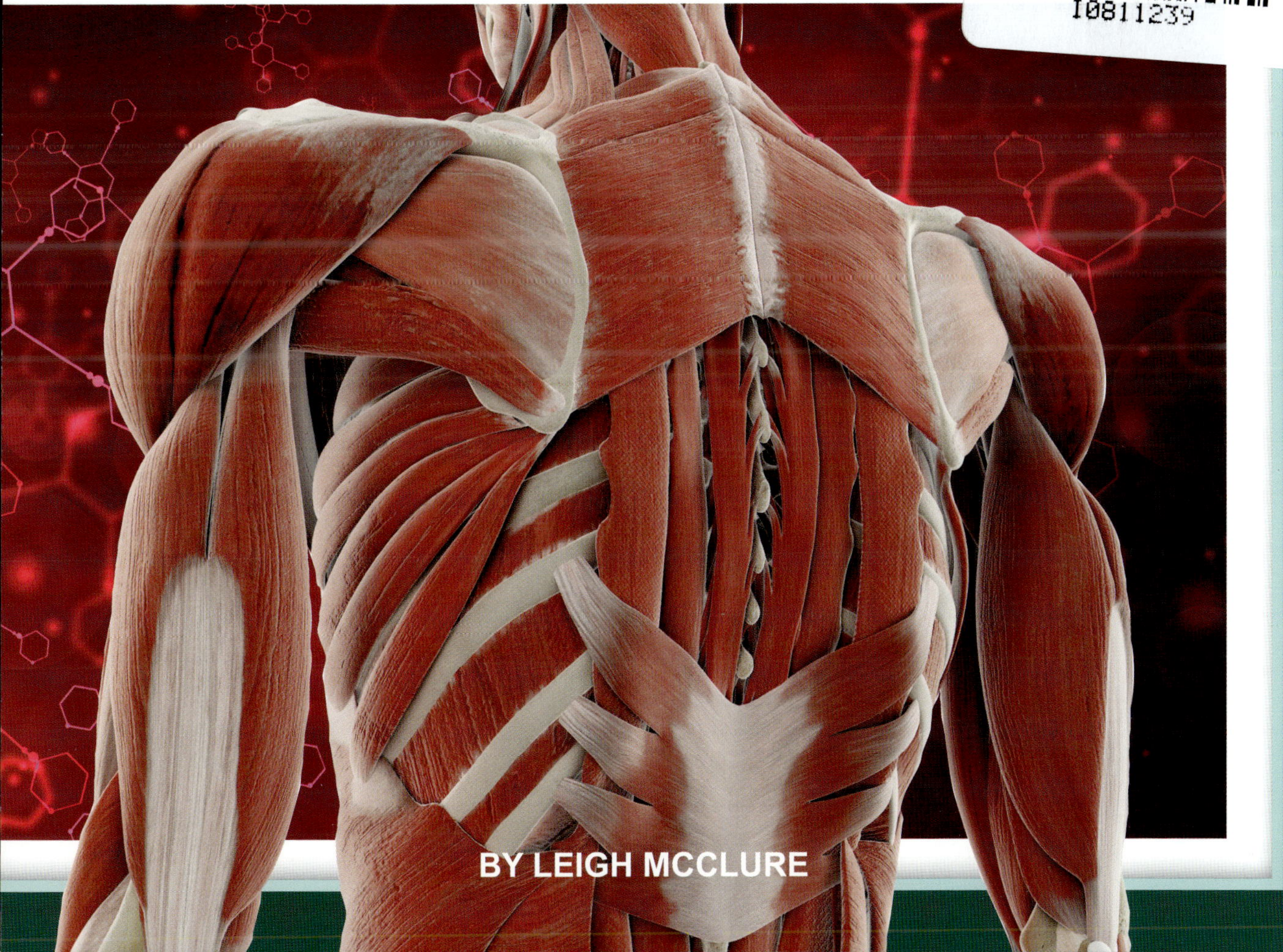

BY LEIGH MCCLURE

Published in 2025 by The Rosen Publishing Group
in association with Scientific American Educational Publishing
2544 Clinton Street, Buffalo NY 14224

Library of Congress Cataloging-in-Publication Data
Library of Congress Cataloging-in-Publication Data Names: McClure, Leigh, author. Title: The muscular system / Leigh McClure. Description: Buffalo, NY : Scientific American Educational Publishing, [2025] | Series: Scientific American investigates the human body | Includes bibliographical references and index. Identifiers: LCCN 2024011205 (print) | LCCN 2024011206 (ebook) | ISBN 9781725351424 (library binding) | ISBN 9781725351417 (paperback) | ISBN 9781725351431 (ebook) Subjects: LCSH: Muscles–Juvenile literature. | Musculoskeletal system–Juvenile literature. | Human anatomy–Juvenile literature. Classification: LCC QM151 .M38 2025 (print) | LCC QM151 (ebook) | DDC 611/.73–dc23/eng/20240324
LC record available at https://lccn.loc.gov/2024011205 LC ebook record available at https://lccn.loc.gov/2024011206

Designer: Andrea Davison-Bartolotta
Editor: Caitie McAneney

Portions of this work were originally authored by Greg Roza and published as *The Muscular System*. All new material in this edition authored by Leigh McClure.

Photo credits: Cover, p. 1 (main) SciePro/Shutterstock.com; cover, p. 1 (background) jijomathaidesigners/Shutterstock.com; p. 5 Brocreative/Shutterstock.com; p. 7 (top) Peter Hermes Furian/Shutterstock.com; p. 7 (bottom) Maridav/Shutterstock.com; p. 9 TarikVision/Shutterstock.com; p. 11 inspiring.team/Shutterstock.com; pp. 13, 15 Sakurra/Shutterstock.com; p. 17 Ground Picture/Shutterstock.com; p. 19 LightField Studios/Shutterstock.com; p. 20 Marcos Castillo/Shutterstock.com; p. 21 Gregory Johnston/Shutterstock.com; p. 22 mariokinhed/Shutterstock.com; p. 23 Tony Stock/Shutterstock.com; p. 25 Pixel-Shot/Shutterstock.com; p. 27 24K-Production/Shutterstock.com; p. 29 VGstockstudio/Shutterstock.com.

Printed in the United States of America

CPSIA compliance information: Batch #CSSA25. For Further Information contact Rosen Publishing at 1-800-237-9932.

CONTENTS

Words in the glossary appear in **bold** type the first time they are used in the text.

LET'S MOVE!

What's your favorite way to move? Maybe you like to play soccer or basketball. Maybe you like to run or climb. All of these activities involve the muscular system. Everything you do, from getting dressed to eating lunch, is all thanks to muscles, whether you're aware of them or not!

Some muscles are always working, even when we aren't thinking about them. These muscles make your heart beat and keep you breathing. Other muscles are controlled by your thoughts, such as when you give a friend a high five. The human body has hundreds of special muscles that carry out important functions, or jobs. In fact, there are around 650 skeletal muscles alone. Let's learn more about this amazing body system and how it helps us every day.

FUN FACT

SKELETAL MUSCLES ARE THE MUSCLES ATTACHED TO BONES.

Running involves several large muscle groups in your body at once.

Building Muscle

Some people build their muscles on purpose. Bodybuilders increase the size of their muscles by lifting weights regularly and eating a special diet. Lifting weights damages, or harms, the cells of the muscles. However, the muscles grow larger and stronger as they heal, especially with the right diet. Protein is a **nutrient** found in many foods that helps repair damaged muscle cells. Bodybuilders need lots of protein.

SUPER SKELETAL MUSCLES

Skeletal muscles are **complex** body parts made up of increasingly smaller bundles of long, **cylindrical** tissues. They include your shoulder muscles and abdominals, or abs. A single muscle is made up of cylindrical parts called fascicles. The fascicles are bundled together inside a covering called the epimysium. A single fascicle contains numerous muscle cells, which are called fibers. Between all muscle fibers and fascicles are connective tissues that hold the cylindrical tissues in place. Blood vessels and nerves run through the connective tissues.

Similar to the fascicles, muscle fibers contain bundles of cylindrical proteins called myofibrils. Inside the myofibrils are thick and thin **filaments**, also called myofilaments, that can slide past each other. Muscles can contract, or grow shorter, because of this sliding.

FUN FACT

SKELETAL MUSCLES ARE STRIATED MUSCLES. THEY HAVE LIGHT AND DARK STRIATIONS, OR STRIPES, WHEN VIEWED UNDER A MICROSCOPE.

Skeletal Muscles

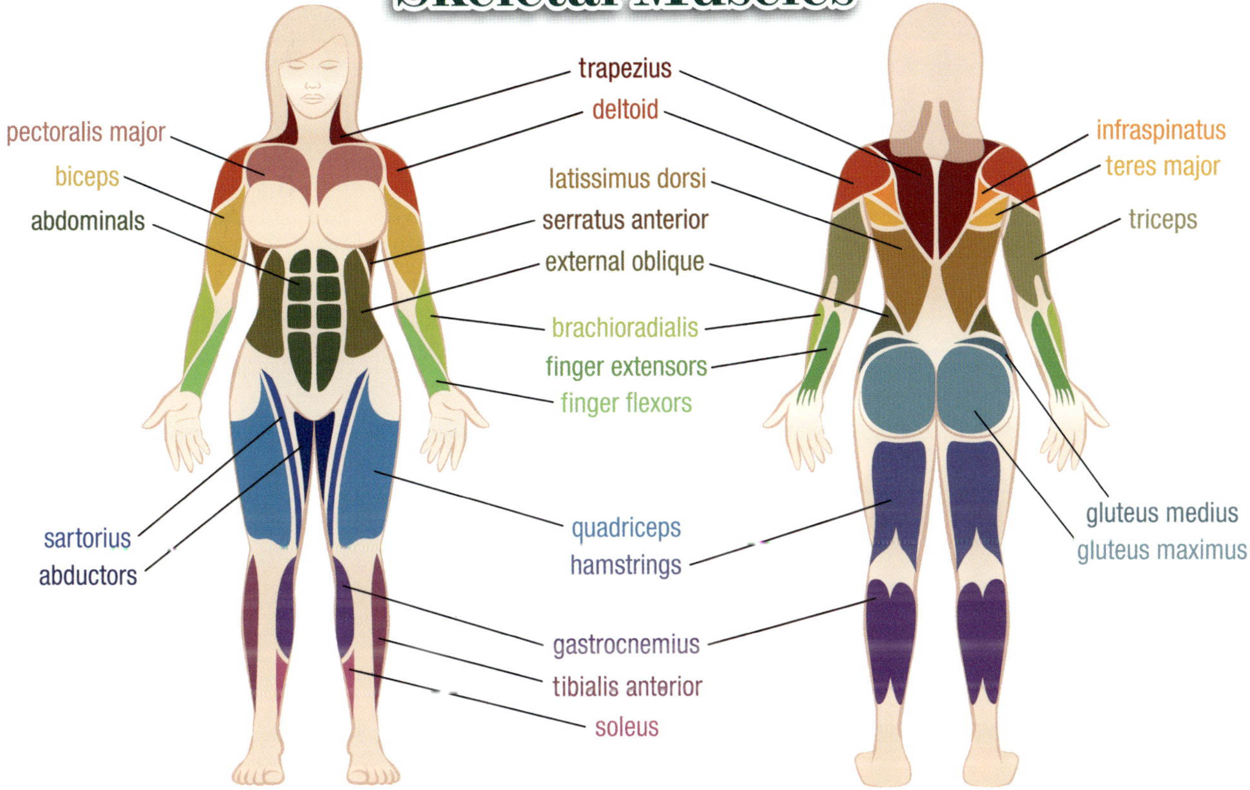

This diagram shows the major muscles that allow you to move every day.

Tough Tendons

Where muscles meet bones, they taper down to tough yet flexible cords called tendons. Tendons transfer motion from muscles to bones. They grow into the bones to form a strong connection that's very hard to break. The largest tendons in the body are the Achilles tendons. They attach the heels to the back of the calves. They're thick and strong because they're used so often.

MAKE IT HAPPEN

Skeletal muscles are different from the other muscles of the body. That's because when it comes to moving them, *you* can make it happen. They are moved voluntarily, or by thinking about it. When you want to move your body, electrical **impulses** travel from your brain, down your spinal column, through your nerves, and to the cells of your skeletal muscles.

Between the nerve cells and muscle cells is a very tiny gap. The electrical impulses tell the nerve cells to release special chemicals that cross the gap and enter the muscle cells. The chemicals tell the myofilaments inside the muscle cells to slide past each other. The cells—which are normally long and cylindrical—become shorter and fatter. The muscle tightens and makes movement.

FUN FACT

A TWITCH IS A QUICK CONTRACTION. FOR EXAMPLE, MUSCLE TWITCHES CAUSE YOUR EYE TO BLINK, WHICH TAKES A FRACTION OF A SECOND.

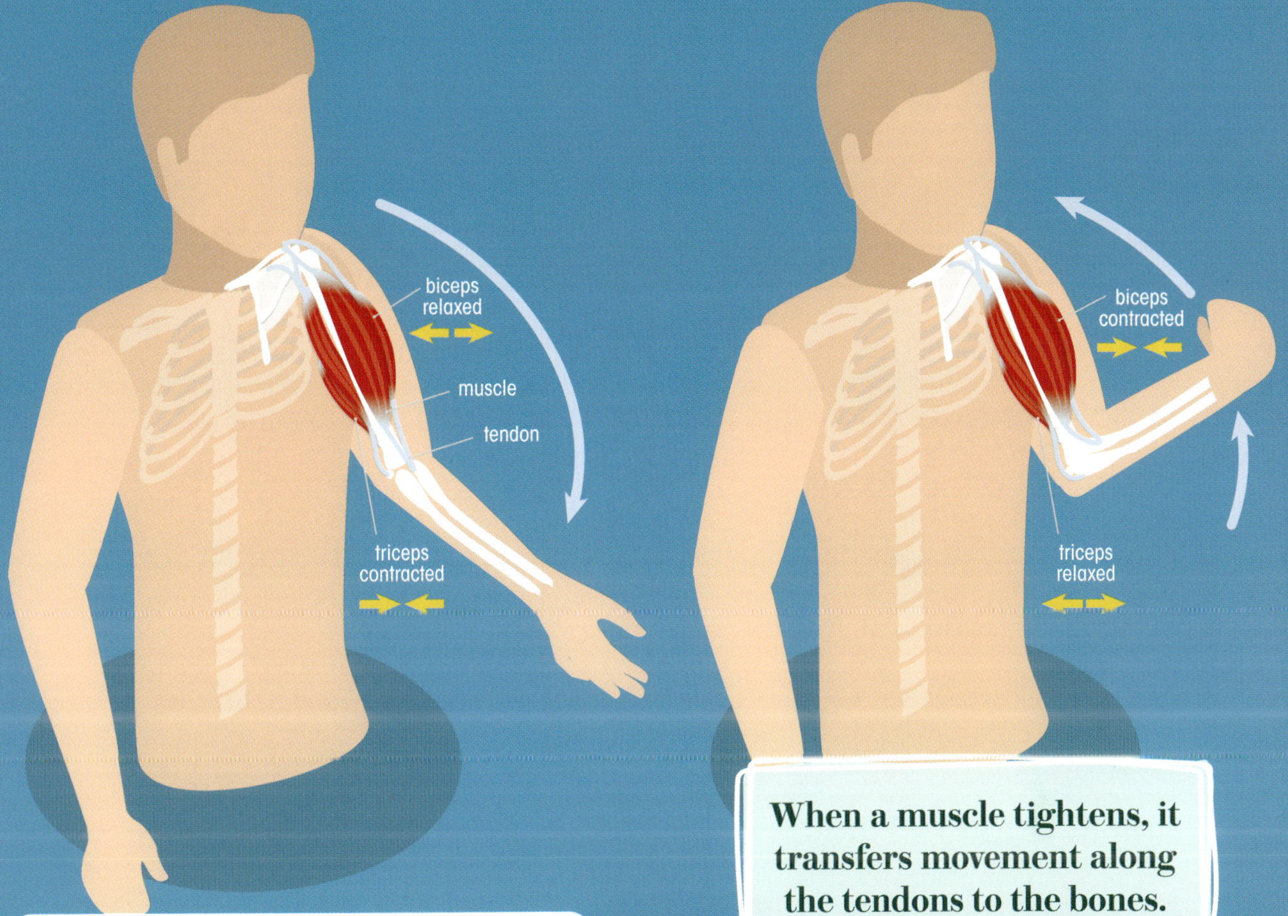

When a muscle tightens, it transfers movement along the tendons to the bones.

Muscle Contractions

Muscle cells are either "off" (relaxed) or "on" (contracted). However, the amount of time the cell remains contracted varies. A quick, single contraction is called a twitch. When a series of twitches occur one after another, the muscle cells remain contracted for a longer period of time. This sustained contraction is called tetanus. Tetanus is also the name of an illness caused by a kind of **bacteria**. It causes the muscles to contract uncontrollably.

EXTENSORS AND FLEXORS

How can muscles move in two directions? They come in pairs of extensors and flexors. These muscles create a push/pull effect to allow our body parts to move in two directions. An extensor extends, or straightens, a joint. A flexor flexes, or bends, a joint.

On the back of your upper arm is an extensor muscle called the triceps. When this muscle contracts, it causes your arm to extend at the elbow and straighten out. Once your arm is extended, the triceps muscle relaxes. A muscle called the biceps is on the front of your arm. This muscle is a flexor. When the biceps muscle contracts, your elbow flexes, or bends. Once your elbow is bent, the biceps relaxes. Try bending your elbow to see these muscles in action!

This stretch will help you stretch your hip flexor muscles. These muscles are near the front of the hips and thighs.

Working Together

Muscles called abductors and adductors work together the same way the extensors and flexors do. Imagine a straight line running from the top of your head to the floor between your feet. Any muscle that moves a body part away from that line is an abductor, such as the deltoid muscles of your shoulders. Any muscle that moves a body part toward the line is an adductor. One example is latissimi dorsi, or the "lats," which are the largest muscles of your back.

ALL ABOUT THE HEART

Another kind of muscle is cardiac muscle. This is the kind that makes up much of the heart. Cardiac muscle makes up the largest layer of the heart walls, known as the myocardium. Cardiac muscle is striated. The actions that cause cardiac muscle cells to contract are nearly the same as those in skeletal muscle cells. However, cardiac muscles are much shorter. They branch out and connect together in a complex network.

Individual cardiac muscle cells aren't told to contract by electrical impulses from the brain. The electrical signals are created within the myocardium itself. Special cardiac cells—called pacemaker cells—use chemical reactions to start the signals, which then travel from cell to cell throughout the heart. Cardiac muscle can contract and relax more than 3 billion times in a single lifetime!

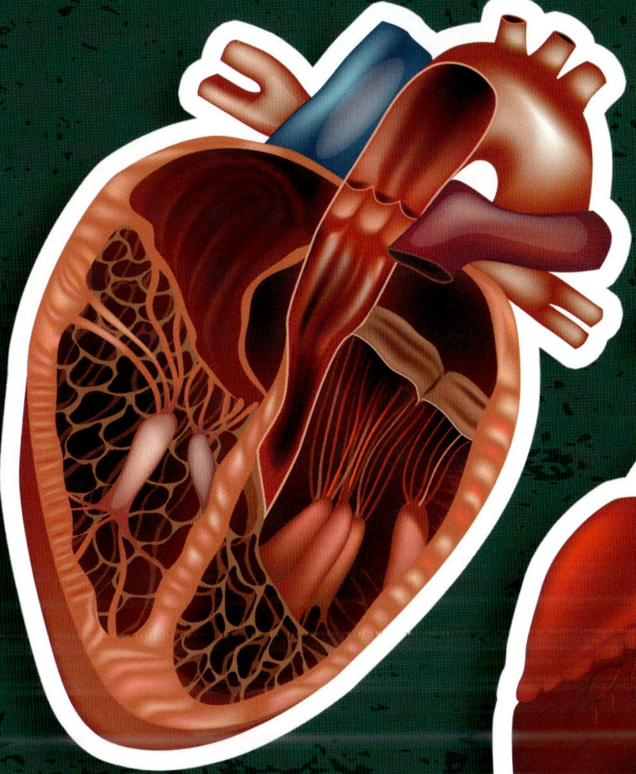

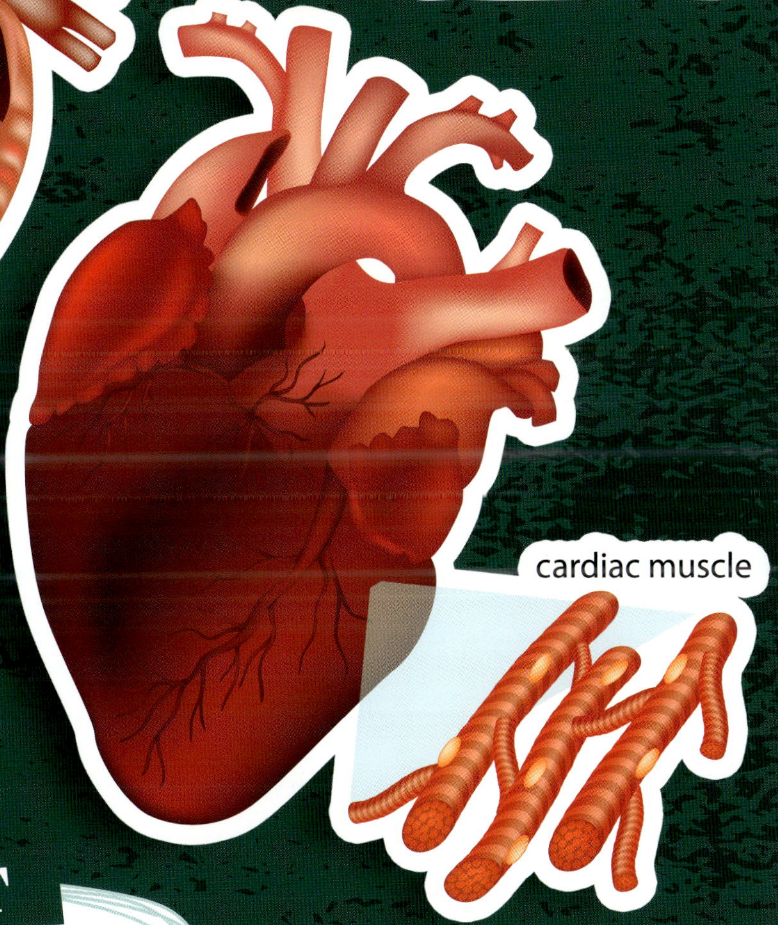

This image shows the anatomy, or makeup, of the human heart.

FUN FACT

CARDIAC MUSCLE MOVES INVOLUNTARILY. THAT MEANS WE DON'T HAVE TO THINK ABOUT MOVING OUR CARDIAC MUSCLE TO KEEP OUR HEART BEATING.

A LOOK AT SMOOTH MUSCLE

Another kind of muscle that moves involuntarily is smooth muscle. It forms the inner surfaces of many hollow organs in the body. Smooth muscle cells are much smaller than skeletal muscle cells. They're wide in the middle and tapered on the ends. Many cells fit tightly together to form thin sheets. The sheets lie one on top of another to form thick layers of muscle.

Just like the other two types of muscle, smooth muscle cells contain bundles of thick and thin filaments that move past each other during contraction. However, smooth muscle also has additional filaments that crisscross the cell and hold the bundles together. When a smooth muscle cell contracts, the additional filaments tighten and squeeze the cells together. These muscles help keep many body systems going, such as the digestive system.

Different Kinds of Muscles

This is what the different kinds of muscle tissue look like. Notice that smooth muscle doesn't have striations.

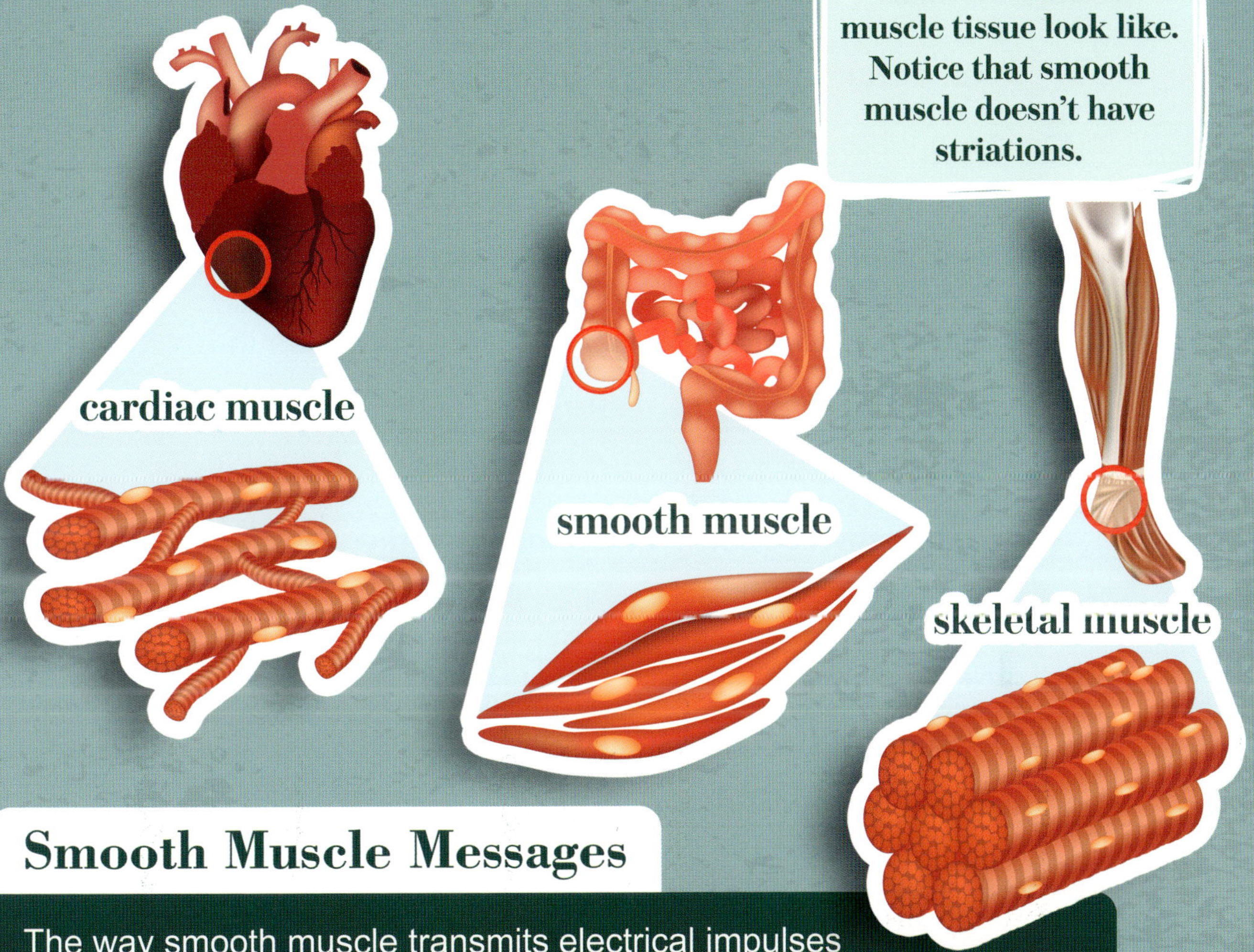

Smooth Muscle Messages

The way smooth muscle transmits electrical impulses varies. Some smooth muscle reacts to **hormones** in the blood. Some has pacemaker cells to initiate electrical impulses. Smooth muscle cells relay electrical impulses similar to the way cardiac muscle cells do. Chemical messengers pass through the boundary between cells. Although all smooth muscle is involuntary, some receives electrical impulses from the brain by way of the nervous system.

Think about all your body does in a day. It's doing things even when you're at rest. Many of these functions are possible because of smooth muscle. Smooth muscle helps in the transportation of nutrients, chemicals, and waste products throughout the body. For example, smooth muscles in an organ called the urinary **bladder** contract to push liquid waste, or urine, from the body when we go to the bathroom.

The digestive system—which includes the esophagus, stomach, and intestines—is lined with smooth muscle. It keeps food and waste products moving. The lower esophagus has two sets of smooth muscle. These smooth muscles work together to move food down to the stomach. This is called peristalsis. Without smooth muscle, peristalsis wouldn't be possible, and we wouldn't be able to eat solid food.

FUN FACT

SMOOTH MUSCLE CAN EVEN BE FOUND IN THE WALLS OF BLOOD VESSELS, WHICH ARE THE PARTS OF THE CIRCULATORY SYSTEM THAT TRANSPORT BLOOD. IT HELPS KEEP THE BLOOD PUMPING PROPERLY.

Take a bite of something, chew, and swallow. See if you can notice peristalsis in the esophagus.

Two Kinds of Smooth Muscle

There are two kinds of smooth muscle. Single-unit smooth muscle is the tightly packed form discussed so far. Multiunit smooth muscle, the less common form, is made up of cells that act independently of each other. Each cell connects to a nerve cell so the body can tell it how to act in different situations. Multiunit smooth muscle can be found in the eye. It causes the iris to grow larger or smaller to let in more or less light.

EXPLORING EXERCISE

Everything you do requires muscles. Smiling, breathing, and eating may seem easy, but some muscles are always at work. Some activities use more muscles than others, and that can make you stronger. For example, when you play a sport or take a walk, you use your muscles to make your whole body healthier. Your muscles grow and your body is better able to prevent illness.

During exercise, the body changes the way it functions to help the muscles work. The lungs take in more oxygen, and the heart rate increases to get more blood to the muscles. You can try an experiment next time you exercise. Notice your breathing and heart rate before you start. Then do 20 jumping jacks. How has your breathing and heart rate changed? How do your muscles feel?

The muscles need oxygen to break down **glucose**, a substance in food. That creates adenosine triphosphate (ATP) which carries the energy our muscles need to do work—like this push-up.

Essential ATP

ATP is present in every cell of the body. It fuels our muscles, but it also provides the rest of our body with the energy necessary to keep us alive. It fuels the activity inside all the cells of the body. ATP also transports substances across **membranes** in the body. It helps nutrients get where they need to go. The process of combining oxygen with glucose to make ATP is called cellular respiration.

Think of how much energy it takes to run a mile. Active muscles burn fuel to create power and movement. The muscles contain enough ATP for quick bursts of activity. Then they need to make more. The muscles have more than one method of doing this.

Sustained exercise includes running, swimming, and biking. When we do these things, we breathe deeper and faster. Our heart pumps faster to make sure our muscles have enough oxygen to keep on creating ATP. This type of muscle activity is called aerobic exercise. "Aerobic" means "with oxygen." Athletes who require great **endurance**—such as marathon runners—participate in daily aerobic workouts. This gets their muscles ready for the major endurance activity of running 26.2 miles (42.2 km)—a whole marathon!

Burning ATP releases a lot of heat, which is why we get hot and sweaty during a workout.

The Benefits of Cardio

Aerobic exercises are often called cardiovascular training, or "cardio." This refers to the cardiovascular system. The cardiovascular system transports oxygen to the muscles and the rest of the body. The main parts of this system are the heart, blood vessels, and blood. However, it also includes the lungs. Regular aerobic exercise strengthens the cardiovascular system. It keeps blood pressure at a healthy level, improves endurance, and builds strong muscles. Runners, cyclists, kickboxers, swimmers, mountain climbers, and dancers use cardio exercises to perform at their best.

What happens when the muscles run out of oxygen to break down glucose? In this case, the glucose instead breaks down into carbon dioxide and a chemical called lactic acid. Lactic acid is what makes our muscles sore during and after exercise. However, muscles also use it as fuel to continue working. This type of muscle activity is called anaerobic exercise. "Anaerobic" means "without oxygen."

Anaerobic muscle activity happens during hard bursts of exercise, such as sprinting, you can only do for a very short time. Strength training is a form of anaerobic exercise. This activity creates lactic acid, which makes the muscles sore. Your muscles ache while this is happening, but they also grow larger and stronger.

When lifting weights, our muscles use quick bursts of energy that don't need to be sustained.

FUN FACT

IT'S IMPORTANT TO REST AFTER ANAEROBIC EXERCISE. THAT GIVES YOUR MUSCLES A CHANCE TO HEAL AND REPAIR.

IN CASE OF INJURY

Sometimes people injure their muscles. This can happen because of inactivity and illness. It can also happen because of overuse.

A cramp is an involuntary contraction of a muscle that won't relax. Cramps have many causes, including muscle fatigue, injuries, inactivity, and a lack of vitamins and nutrients.

A strain, or pulled muscle, is a common muscle injury. Normally, muscle fiber is tightly woven. In a strain, the fibers become overstretched or torn. The muscle usually repairs itself within a few weeks. Some strains, however, can last a long time when not properly cared for. A rupture is a more serious strain. This happens when the muscle is completely torn. Some serious muscle tears can only be fixed by surgery.

FUN FACT

A HEMATOMA IS A DEEP BRUISE. AFTER A FALL OR A HARD BLOW TO A MUSCLE, BLOOD VESSELS CAN BREAK. THIS CAUSES SWELLING AND A BLOOD CLOT IN THE MUSCLE.

Sometimes people need to go to physical therapy to get better after a muscle injury.

Remember to RICE

Being physically fit can help you avoid muscle injuries. However, they can happen to anybody. When caring for a strained muscle, remember the following word: RICE. That stands for Rest, Ice, **Compression**, and Elevation. Also, always ask a doctor for advice about muscle injuries. To avoid muscle injuries, do some light cardio before exercising to get blood flowing to the muscles. Follow the cardio by stretching the muscles you intend to work out. Stretching after a workout may help prevent injury as well.

DISEASES AND DISORDERS

Myopathy is another word for a disease of the muscles. Myopathies are often caused by the breakdown of muscle fibers. Depending on the specific illness, symptoms may include pain, weakness, cramps, stiffness, bruising, and swelling. Some myopathies are **hereditary**, while others aren't related to family history. Some muscle disorders are the result of other illnesses. They range from mild to severe.

Some of the most life-threatening myopathies are more than 30 hereditary illnesses called muscular dystrophies (MD). These diseases are marked by weakness and death of skeletal muscles. Some forms affect the cardiac muscles.

Neuromuscular diseases are illnesses that affect the nerves that control skeletal muscles. They include Parkinson's disease and multiple sclerosis. English scientist and writer Stephen Hawking (1942–2018) had a neuromuscular disease called amyotrophic lateral sclerosis (ALS).

FUN FACT

MUSCULAR DYSTROPHY CAN BE MILD AND PROGRESS SLOWLY. HOWEVER, SOME KINDS ARE SEVERE AND FAST ACTING. THERE'S NO CURE, BUT THERE ARE TREATMENTS THAT CAN HELP.

Some people with muscular dystrophy get around using wheelchairs.

Cardiac Muscle Disease

Myocarditis is a swelling of the myocardium, the cardiac muscle. Most often this occurs as a result of a viral infection. Signs of myocarditis include chest pain, abnormal heartbeat, difficulty breathing, fever, swelling in the ankles and feet, and fatigue. This illness is treated with rest and heart medicines. More serious cases may require surgery. If left untreated, myocarditis can result in heart failure, heart attack, stroke, and sudden death.

KEEP YOUR MUSCLES HEALTHY!

How can you treat your muscles right? Give them fuel, exercise, and rest. You can get your body moving by taking walks, runs, bike rides, and swims. Then, you can give your body fuel by eating healthy **carbohydrates**, such as whole grains, vegetables, and fruits. These provide glucose for ATP. Protein is important too. Eat plenty of fish, chicken, beans, and nuts. This will help your muscles repair and grow stronger. Drink lots of water when you work out too. When your body needs to heal, give it the rest it needs.

The muscular system helps you get up and around. It keeps your body moving. Next time you do your favorite activity, think of all the muscles that make it happen!

It takes about 47 muscles to frown and only 13 to smile!

FUN FACT

ABOUT 40 PERCENT OF A PERSON'S WEIGHT IS MADE UP OF MUSCLE.

GLOSSARY

bacteria: Tiny creatures that can only be seen with a microscope.

bladder: An expandable organ that stores liquid waste until it can be removed from the body.

carbohydrate: A nutrient in many types of food that the body uses as a source of energy.

complex: Having to do with something with many parts that work together.

compression: The act of pressing or squeezing together.

cylindrical: Shaped like a cylinder or tube.

endurance: The ability to do something for a long time without getting tired.

filament: A slender strand of a material.

glucose: A sugar the body uses for energy.

hereditary: Passed from parent to child.

hormone: A chemical made in the body that tells another part of the body how to function.

impulse: A force that causes motion.

membrane: A thin tissue in the body.

nutrient: Something a living thing needs to grow and stay alive.

FOR MORE INFORMATION

Books

Rea, Amy C. *The Muscular System*. Minneapolis, MN: Kids Core, 2023.

Vizcarra, Natasha. *The Muscular System*. New York, NY: Children's Press, 2024.

Walker, Richard. *Human Body*. New York, NY: DK Publishing, 2023.

Websites

Biology for Kids: Human Body
www.ducksters.com/science/biology/humanbody.php
Learn more about the different body systems, including the muscular system.

15 Facts About the Human Body!
www.natgeokids.com/uk/discover/science/general-science/15-facts-about-the-human-body/
Explore 15 more fun facts about the human body with National Geographic Kids.

Your Muscles (For Kids)
kidshealth.org/en/kids/muscles.html
Check out interesting facts about the muscular system.

INDEX

Cuddly Koalas

Growing Up Wild

Madison Parker

Norwood House Press

Cataloging-in-Publication Data

Names: Parker, Madison.
Title: Cuddly koalas / Madison Parker.
Description: Buffalo, NY : Norwood House Press, 2026. | Series: Growing up wild | Includes glossary and index.
Identifiers: ISBN 9781978573994 (pbk.) | ISBN 9781978574007 (library bound) | ISBN 9781978574014 (ebook)
Subjects: LCSH: Koala--Juvenile literature. | Koala--Infancy--Juvenile literature.
Classification: LCC QL737.M384 P375 2026 | DDC 599.2'5--dc23

Published in 2026 by
Norwood House Press
2544 Clinton Street
Buffalo, NY 14224

Designer: Rhea Magaro
Editor: Kim Thompson

Photo credits: Cover, p. 1 slowmotiongli/Shutterstock.com; p. 5 Carla Thomas/Shutterstock.com; p. 6 thanongsuk harakunno/Shutterstock.com; p. 7 Susan Flashman/Shutterstock.com; p. 8 BMCL/Shutterstock.com; p. 10 Damian Kuzdak/Shutterstock.com; p. 11 TRossJones/Shutterstock.com; p. 13 EQRoy/Shutterstock.com; p. 14 Yori Hirokawa/Shutterstock.com; p. 15 nattanan726/Shutterstock.com; p. 17 Alizada Studios/Shutterstock.com; p. 18 Andreas Deak/Shutterstock.com; p. 19 Chirandilshan/Shutterstock.com; p. 21 Wayne Buttersworth/Shutterstock.com;

Printed in the United States of America

CPSIA compliance information: Batch #CSNHP26: For further information contact Norwood House Press at 1-800-237-9932.